Understanding
Topical Corticosteroids

WHAT YOU NEED TO KNOW

By Jerome Shupack, MD

Dedication

This book is dedicated to Dr. Frances Pascher, an inspired teacher and clinician who helped launch my career in Dermatology.

Table of Contents

Introduction

Since the 1950s, topical corticosteroids have been considered a cornerstone in dermatological therapeutics. Topical corticosteroids are still widely used today in the treatment of skin diseases. They are not used to cure disease, but they can effectively control the symptoms. However, their widespread misuse has led to considerable side effects. Although topical corticosteroids are generally well tolerated, they can cause skin atrophy and less frequently cause hypopigmentation, secondary infections, and acne. The guiding principles of treatment are to use a steroid of suitable potency for the area of lesions to be treated, to apply the product sparingly, and to limit the use of topical steroids to the shortest time needed to obtain a clinically acceptable effect. Topical corticosteroids can be absorbed through the skin, and the risk of systemic absorption increases with long periods of application and if occlusive dressings are used. To understand the pharmacology and pharmacokinetics of topical corticosteroids, every physician must first understand how their clinical effect is measured and compared.

Classification of Topical Steroids

Topical corticosteroid preparations vary greatly in potency. Classification depends on many factors, including the drug's characteristics and concentration as well as the vehicle or base in which it is delivered. The term potency is used to describe the intensity of a topical corticosteroid's clinical effects. High and super high potency topical steroid preparations should generally be avoided on the face, armpits, groins, and anal folds, because they may cause significant skin thinning and other complications. The skin in these situations is thin with a proportionally larger surface area and susceptible to atrophy as well as systemic absorption.

Vasoconstrictor Assay

The vasoconstrictor assay is the most commonly used test to determine the clinical effects of topical corticosteroids. In 1961 a publication appeared that indicated that topical corticosteroids induce blanching or vasoconstriction of normal skin when applied topically. This turned out to be a very useful screening procedure to sort out the hundreds of corticosteroids available and to determine whether or not they would be useful for topical application in cutaneous disease. This assay has been proven to predict accurately comparative clinical effectiveness, and to determine which formulations of any given steroid will be the most powerful. Steroids can be rendered more or less potent depending upon which formulation incorporates them. It is possible that one corticosteroid can have four different categories of activity, depending on its formulation, at the same concentration. The pharmacokinetics and resultant clinical potency of a topical corticosteroid preparation depend on three main interrelated factors: the structure of the corticosteroid molecule, the vehicle, and the skin onto which it is applied. The vehicle can indirectly alter a given preparation's therapeutic and adverse actions by altering the pharmacokinetics of the topical corticosteroid molecule.

2

Formulation and Relation to Potency

It is well known that a given steroid at a given concentration will give a different biologic potency when incorporated into an ointment as compared to a conventional cream. The ointment tends to enhance to biologic activity of almost all of the corticosteroids. The addition of propylene glycol in any vehicle tends to enhance penetration of a given corticosteroid.

Each corticosteroid is different in its response to formulation and very small changes in the basic formulation may make rather major changes in the ability of the corticosteroid to penetrate and be biologically active. It is possible to dilute the given commercial corticosteroid formulations with vehicles other than those recommended by the manufacturer. The potency of a corticosteroid formulation can be readily decreased by diluting it with a foreign vehicle. There is also the problem of stability of a steroid. Some foreign vehicles will destabilize the corticosteroid so that it will break down over a period of time and render the formulation ineffective.

The occlusion of the skin surfaces can result in a 10 to 100-fold increase in biologic activity of a given corticosteroid. Occlusion can so enhance penetrability that it can overcome any difference in the basic formulation of the steroid. In order for occlusion to affect this great increase in penetration, it must be present for over a 7-8 hour period, and preferably 12-24 hours. Occlusion of standard formulations is sometimes used in the management of certain dermatoses, but it is cumbersome and uncomfortable. The steroid can be incorporated in tape for occlusive application as in a product called Cordran® tape (Watson Pharmaceuticals) that has flurandrenolone acetonide incorporated into a semi-clear tape that comes in the form of 4 mcg/cm² patches or rolls.

REGIONAL DIFFERENCES IN PENETRATION

The primary layer that causes resistance to penetration is the stratum corneum. Penetration of the applied drug correlates inversely with the thickness of the stratum corneum in any given area. The condition of the skin also affects bioavailability. Penetration increases with inflamed or diseased skin and also with increased hydration of the stratum corneum, relative humidity, and temperature. For example, chemical agents may pass quantitatively through the mucous membrane if they are held in place long enough. When treating the palms and soles, higher dosage and potency may be required because the skin is thicker. Penetration through the nails is generally so small that therapeutic levels of corticosteroids cannot be passed through the nails into the nail bed. With newer super high potency preparations, and with non-porous tape, occlusion, and a willing and reliable patient, it may be possible to drive enough corticosteroid into the nail bed to achieve a therapeutic effect close to that obtained with intralesional injections. The stratum corneum may also act as a reservoir for topical steroids for up to five days; this retention is concentration and formulation dependent. Some skin diseases cause damage to the stratum corneum, which results in an increase in penetration of corticosteroids. As the skin improves, the stratum corneum regenerates itself, and as it becomes thicker, there is less penetration of the corticosteroid.

TABLE I

REGIONAL DIFFERENCES IN PENETRATION — FROM MOST TO LEAST

1. Mucous membrane

2. Scrotum

3. Eyelids

4. Face and scalp

5. Chest and back

6. Upper arms and legs

7. Lower arms and legs

8. Dorsa of hands and feet

9. Palmar and plantar skin

10. Nails

TABLE II

TOPICAL STEROID CLASSIFICATION BY POTENCY

LOW

ACTIVE INGREDIENT(S)	FORMULATION	PACKAGE SIZE	PRODUCT
Alclometasone dipropionate cream, ointment	0.05%	15, 45, 60g	Aclovate
Desconide cream, ointment, lotion	0.05%	15, 60g; 2, 4oz	DesOwen
Hydrocorisone acetate & pramoxide HCl cream, ointment, lotion	1%	1, 2oz; 2, 4oz	Pramosone
Hydrocorisone acetate & pramoxide HCl cream, ointment, lotion	2.5%	1, 2oz; 2, 4, 8oz	Pramosone
Hydrocortisone cream	1%	60mL, 30/60g OTC brands	Many

INTERMEDIATE

ACTIVE INGREDIENT(S)	FORMULATION	PACKAGE SIZE	PRODUCT
Fluticasone propionate ointment	0.005%	15, 45, 60g	Cutivate
Fluticasone propionate cream	0.05%	15, 45, 60g	Cutivate
Prednicarbate emollient cream	0.1%	15, 60g	Dermatop
Mometasone furoate cream, lotion	0.1%	15, 45g 30, 60mL	Elocon
Hydrocortisone butyrate cream, ointment, solution	0.1%	15, 45g; 20, 60mL	Locoid
Hydrocortisone butyrate cream	0.1%	15, 45g	Locoid Lipocream

Betamethasone valerate foam	0.12%	100g	Luxiq
Hydrocortisone valerate cream, ointment	0.2%	15, 45, 60g	Westcort

HIGH

ACTIVE INGREDIENT(S)	FORMULATION	PACKAGE SIZE	PRODUCT
Amcinonide cream, ointment, lotion	0.1%	15, 30, 60g; 20, 60 mL	Cyclocort
Augmented betamethasone dipropionate cream	0.05%	15, 50g	DiproleneAF
Fluocinonide cream, gel, ointment, solution	0.05%	5, 30, 60, 120g; 20, 60cc	Lidex
Fluocinonide cream	0.05%	15, 30 60g	Lidex-E
Diflorasone diacetate cream	0.05%	15, 30, 60g	Psorcon
Diflorasone diacetate emollient cream	0.05%	15, 30, 60g	Psorcon-E
Desoximetasone cream, gel, ointment	0.25%	15, 60g	Topicort
Desoximetasone emollient cream	0.05%	15, 60g	Topicort LP
Mometasone furoate ointment		15, 45g; 30, 60mL	Elocon

SUPER HIGH

Active Ingredient(s)	Formulation	Package Size	Product
Augmented betamethasone dipropionate gel, ointment lotion	0.05%	15, 50g; 30, 60mL	Diprolene
Diflorasone diacetate ointmen	0.05%	15, 30, 60g	Psorcon
Clobetasol propionate cream, gel, ointment, scalp appl.	0.5%	15, 30, 45, 60g; 25, 50mL	Temovate
Clobetasol propionate emollient cream	0.05%	15, 30, 60g	Temovate-E
Halobetasol propionate cream, ointment	0.05%	15, 50g	Ultravate

TABLE III

CLASSIFICATION	SITES
Low	Face, genital skin, babies' skin, skin folds
Intermediate	Similar to Low except where there is some thickening of the skin from the dermatoses (i.e., psoriasis with scales)
High	Thick skin lesions
Super High	Palms, soles, elbows, knees, dermatoses characterized by hyperkeratotic lesions

3

General Considerations

For all topical drugs, the physician should choose the active ingredient coupled with an appropriate base that suits the dermatological condition best. A sufficient but not excessive quantity should be prescribed. Cream and lotion basis should vanish completely into the skin after application. The proper administration of topical corticosteroid therapy involves choosing a preparation, estimating the necessary amount, and then supervising the treatment. Ultimately, patients' compliance and clinical response determine the best vehicle; i.e. an ointment base may be theoretically the best preparation but patients may not use the medication because of the greasiness. For products that require application twice daily, it is wise to prescribe a cream or lotion base for the morning, and an ointment base for bedtime use. It is generally considered that 20-30 grams of a topical medicine is sufficient to cover the entire cutaneous surface of an average human once.

TABLE IV
FACTORS INFLUENCING CHOICE OF BASE

1. The area to be treated
2. Potential for irritation
3. History of previous allergic reactions
4. The age of the patient
5. The site to be treated
6. The extent of the disease
7. Occlusion vs. non-occlusion

TABLE V
SITE-BASE CONSIDERATIONS

BASE	SITE
Creams	Weeping skin, body folds and other moist conditions or areas require a cream formula.
Gels	Less greasy than ointments, best for sites and individuals with much natural oil
Ointments	Dry skin conditions, like eczema, xerosis, and psoriasis.
Lotions	Hair-bearing areas like scalp psoriasis.
Foams	Any area, especially scalp, easy application with less grease
Solutions	Hairy areas
Tape	Reserved for localized stubborn areas

TABLE VI

SPECIFIC AND SELECTED APPLICATIONS OF TOPICAL CORTICOSTEROID USE IN DERMATOLOGY

Dermatitis/Papulosquamous

Atopic dermatitis

Diaper dermatitis

Dyshidrotic eczema

Erythroderma

Lichen simplex chronicus

Nummular dermatitis

Pityriasis rosea

Psoriasis – intertriginous

Psoriasis – plaque or palmoplantar

Seborrheic dermatitis

Bullous dermatoses

Bullous pemphigoid

Cicatracial pemphigoid

Epidermolysis bullosa acquista

Herpes gestationis

Pemphigus foliaceus

Other uses

Alopecia areata

Acne keloidalis nuche

Chondrodermatitis nodularis helicis

Cutaneous T-cell lymphoma, patch-stage

Granuloma annulare

Jessner's lymphocytic infiltrate

Lichen planopilaris

Lichen sclerosis et atrophicus

Morphea

Prurutus urticarial papules and plaques of pregnancy

Pruritus – perianal, vulvar, scrotal

Sarcoidosis

Vitiligo

Wells' syndrome

4

Pharmacology of Topical Corticosteroids

Topical corticosteroids are anti-inflammatory and immunomodulatory substances. The skin, by virtue of its interface role between the inner and outer environments, participates strongly in the body's defense mechanisms. Topical corticosteroids will suppress T-cell activity in the skin induced by a variety of internal or external noxious events and thus exert an anti-inflammatory effect. This explains their widespread use for the treatment of inflammatory dermatoses.

For topical corticosteroids to be effective they must be absorbed into the skin, and issues of absorption have already been discussed. There is some binding or reservoir effect of steroids in the stratum corneum so that when used in areas with absent or minimal stratum corneum (i.e., eyelids, scrotum, lips), more frequent application may be necessary. The epithelial cells of the skin do contain the drug metabolizing microsomal cytochromes so that some local drug metabolism does occur in skin. Once through the epidermis, topical corticosteroids are rapidly picked up by the superficial dermal capillary blood vessels and carried away and handled as small dose systemic corticosteroids. For inflammatory dermatoses involving the mid or lower dermis (i.e., granuloma annulare), it therefore becomes necessary to use a more potent preparation to achieve a significant tissue level before the drug is entirely distributed to the systemic circulation.

Dermatoses involving the subcutaneous fat (i.e. various panniculidides) do not respond to topical corticosteroids and may require intralesional or even systemic therapy.

5

Most Common Skin Disorders by Disease

SUPERFICIAL DERMATITIDES

Atopic Dermatitis

Atopic dermatitis is a chronically relapsing inflammatory skin disorder that is often described as the result of an allergic response of the skin to environmental allergens such as food, inhalant, or cutaneous antigens. It is a troublesome condition commonly seen in primary care practices. The disease usually begins in early infancy with initial onset during the first year of life. The early eruptions involve the wet areas in children (i.e. the face, diaper area skin). As the child develops, the lesion distribution shifts to the dry and lichenified eruptions involving the antecubital and popliteal fossae, hands, back of neck, and almost anywhere else with intense pruritis.

The underlying pathological mechanisms are not fully understood; however, several factors such as genetic predisposition, climate, and exposure to environmental irritants are suggested to play a role in this complex multifactorial skin disorder.

While most patients with atopic dermatitis have family members with similar problems and/or asthma or upper respiratory allergies, twenty (20) percent may be the only affected family member. Although there is no cure, the condition can be managed through the use of topical corticosteroids, and more recently, several topical non-steroidal immunomodulating preparations have become available.

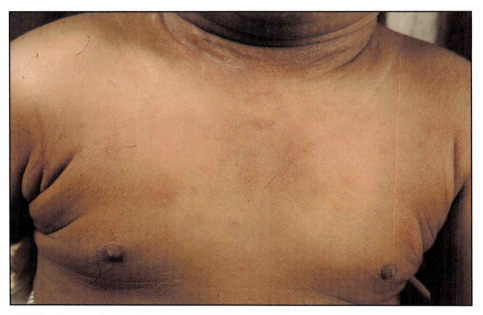

Childhood atopic dermatitis

The course is unpredictable. Although the dermatitis often improves by age 3 or 4 years, exacerbations are common throughout childhood, adolescence, or adulthood. Pruritus is constant; subsequent scratching and rubbing lead to an itch-scratch-rash-itch cycle. The dermatitis may become generalized. Secondary bacterial infections and regional lymphadenitis are common. Atopic dermatitis sufferers are colonized with Staphylococcus aureus and the organism has been isolated from infected eczema, from chronic lesions and from clinically normal skin of patients. Addressing the bacterial component of the disease is often overlooked until the point that oral antibiotics are indicated. When the skin is infected, treating with steroids may permit proliferation of the bacteria, so that concomitant antibacterial topical or systemic treatment is important in the management of many atopic dermatitis patients.

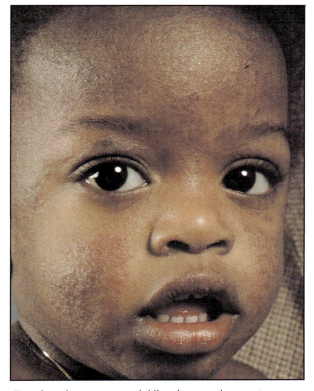

Facial predominance in childhood atopic dermatitis

Contact Dermatitis

Contact dermatitis is an acute or chronic inflammation, often asymmetric or oddly shaped, produced by substances contacting the skin and causing irritant or allergic reactions. Contact dermatitis may be caused by a primary chemical irritant or by an allergen or by sensitization to a specific molecule. Contact dermatitis ranges from transient redness to severe swelling with blistering. Typically, the dermatitis is limited to the site of contact but may later spread. If the causative agent is removed, erythema disappears within a few days to weeks and blisters dry up. Vesicles and bullae may rupture, ooze, and crust. As inflammation subsides, scaling and some temporary thickening of the skin occur. Continued exposure to the causative agent may perpetuate the dermatitis. Typical skin changes and a history of exposure facilitate diagnosis, but confirmation may require exhaustive questioning and extensive patch testing. The

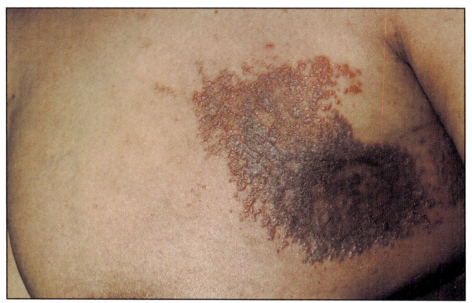

Allergic contact dermatitis

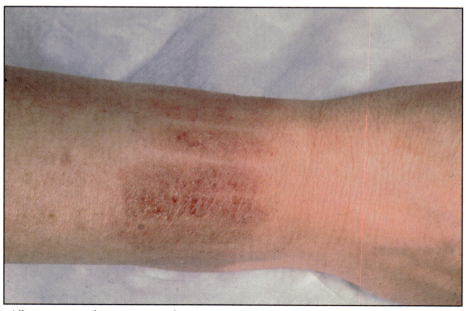

Allergic contact dermatitis secondary to wrist watch

patient's occupation, hobbies, household duties, vacations, clothing, topical drug use, cosmetics, and spouse's activities must be considered. Knowing the characteristics of irritants or topical allergens and the typical distribution of lesions is helpful. The site of the initial lesion can also be an important clue.

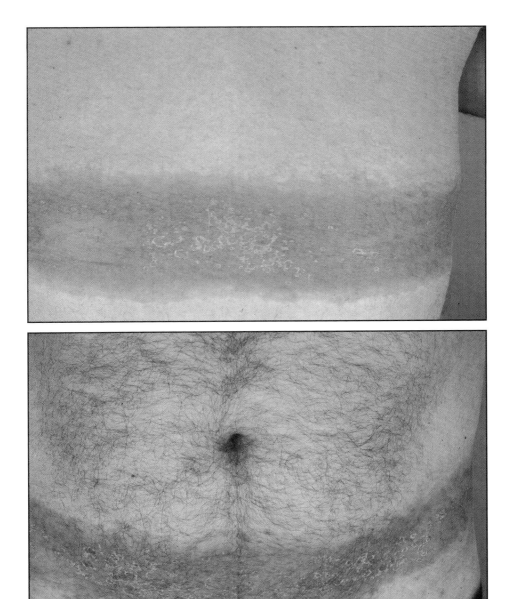

Above two images: Allergic contact dermatitis secondary to waistband. Note the well demarcated nature of the dermatitis, suggesting an external cause

Seborrheic Dermatitis

Seborrheic dermatitis most often occurs in babies younger than 3 months of age and in adults from 30 to 60 years of age. In adults, it is more common in men than in women. It usually affects the scalp, eyebrows, bridge, and side of the nose. Seborrheic dermatitis can also affect the skin on other parts of the body, such as the mid chest and body folds. Seborrheic dermatitis usually causes the skin to look a little greasy, scaly, dry, and irritated.

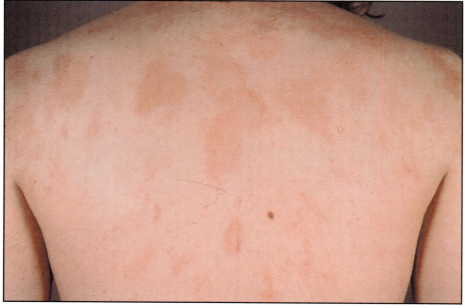

Above: Seborrheic dermatitis of the upper back

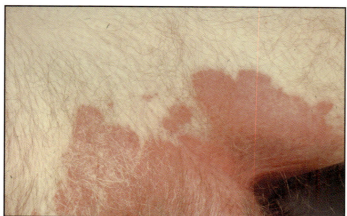

Right: Intertriginous seborrheic dermatitis

Stasis Dermatitis

Stasis dermatitis refers to the skin changes associated with accumulation of fluid in the legs. Varicose veins, congestive heart failure, and other conditions can cause swelling of the extremities, especially the feet and ankles. This swelling (edema) is caused when plasma (the fluid portion of blood) leaks out of the blood vessels and into the tissues. The excess fluid in the tissues interferes with the blood's ability to feed the tissue cells and dispose of cellular waste products. The tissue becomes poorly nourished and fragile, resulting in stasis dermatitis. The disorder is common on the ankles because there is less supportive tissue in this area. The skin becomes thin and inflamed, and open ulcers may form and heal slowly.

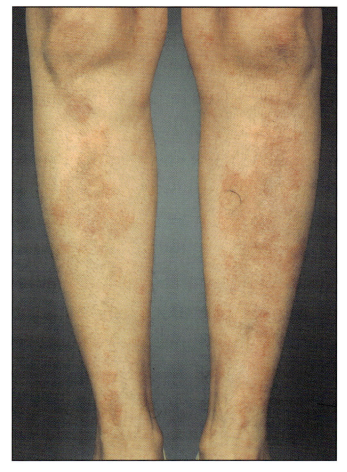

Stasis dermatitis in a common pretibial location

The skin may darken. The skin, initially thin, may later thicken, perhaps because of itching and scratching of the area.

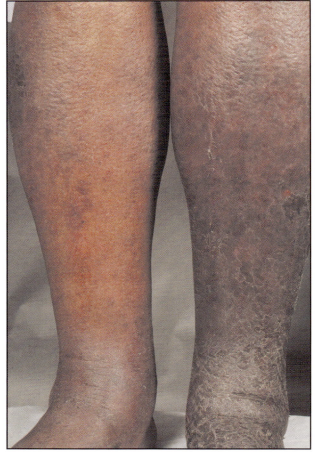

Chronic stasis dermatitis with pronounced lichenification of the left lower extremity

Nummular Dermatitis

This is a pruritic dermatitis of unknown cause, thought to be a skin response to some antigenic stimulus which can only rarely be identified. Hypersensitivity to bacteria on the skin results in crusted patches. Dry skin in the winter months can cause dry, non-itchy, coin-shaped patches. It can affect any part of the body. One or many patches appear, and may persist for weeks or months. The majority of patches are round or oval, coin or disc-shaped dermatitis. They can be several centimeters across, or as small as two millimeters. The skin between the patches is usually normal, but may be dry and irritable. Nummular dermatitis may be extremely itchy, or scarcely noticeable. When the patches clear, they may leave hyper or hypopigmentation for some weeks or months.

Lichen Simplex Chronicus

Lichen simplex chronicus is a troublesome intractable itchy dermatosis, which may persist despite intensive topical treatments. It is characterized by a self-per-

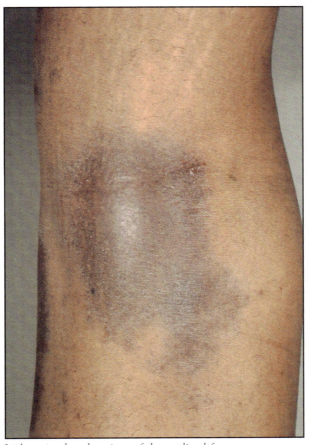

Lichen simplex chronicus of the popliteal fossa

petuating scratch-itch cycle. Lichen simplex chronicus may be a result of something, such as clothing, that rubs or scratches the skin, or irritation of the skin causing the person to rub or scratch an area. This causes thickening of the skin as a response to chronic irritation. The thickened skin itches, causing more scratching, causing more thickening. The skin may become leathery. The disorder may cause brownish pigmentation of the skin in the lesion area. This disorder may be associated with atopic dermatitis or psoriasis. It may also be associated with nervousness, anxiety, depression, and other psychological disorders. It

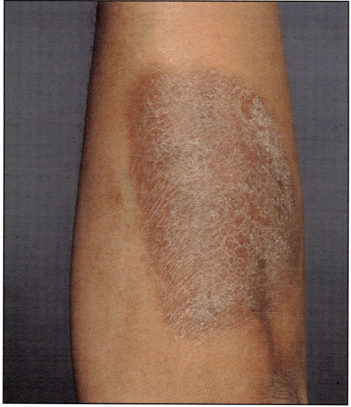

Lichen simplex chronicus of the popliteal fossa

can be common in children, who chronically scratch insect bites and other areas. It can also be common in mentally retarded children who have chronic repetitive movements.

Pityriasis Rosea

Pityriasis rosea is a skin disorder involving a characteristic rash. This disorder is a common skin rash of young people, especially young adults. It occurs most commonly in the fall and spring. Attacks generally last 4 to 8 weeks. The eruption may be pruritic or without symptoms. The episode may disappear by 3 weeks or

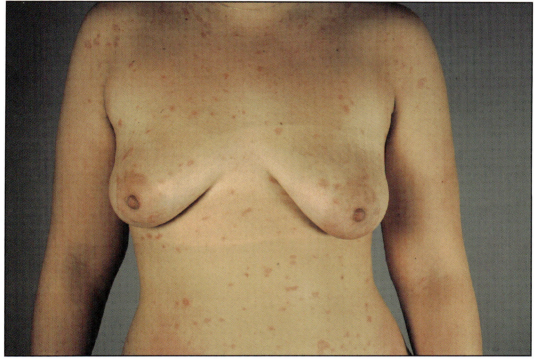

Well circumscribed plaque of lichen simplex chronicus

last as long as 12 weeks. There is generally a single larger patch called a herald patch followed several days later by additional oval patches with a central collarette of scale distributed predominantly on the central trunk and following the pattern of the peripheral spinal nerves. Pityriasis rosea can be diagnosed based on the clinical appearance of the rash. However, a blood test may be useful to distinguish pityriasis rosea from the very similar rash seen in secondary syphilis.

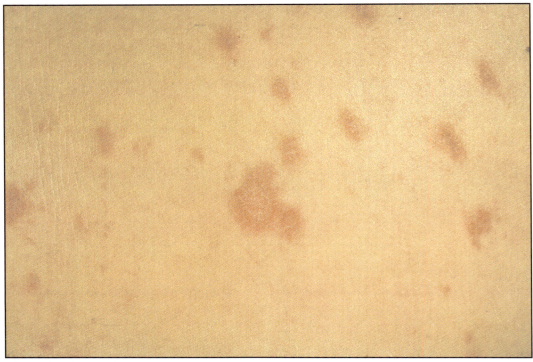

Classic distribution of pityriasis rosea

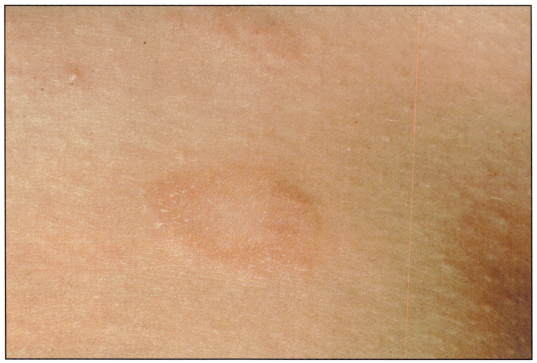

Herald patch of pityriasis rosea

Psoriasis

Psoriasis is the result of a disordered immune system. The T-cells, a type of white blood cell, become over-stimulated and direct the skin to try and heal a non-existent injury. The skin reacts by growing very fast, trying to grow the infection off the skin. These areas become the reddened, inflamed, patches with white scale on them.

The inflammation process of psoriasis is somewhat deeper than the above-mentioned conditions and also involves a thickening of the epidermis, which inhibits the absorption of topical steroids. Topical corticosteroids remain the most commonly prescribed agents for psoriasis. One often needs to occlude the area to be treated to assure adequate penetration, except for areas of skin folds. Despite the

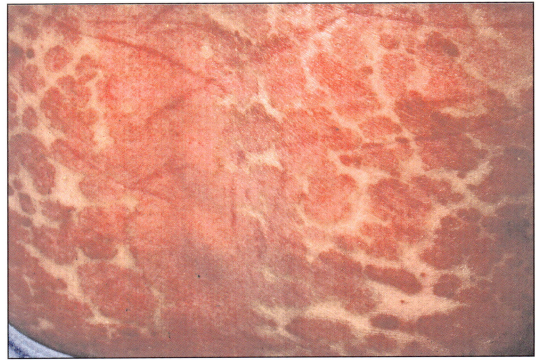

Psoriasis

availability of new treatments for psoriasis, complete clearing of psoriasis is not a realistic expectation from topical monotherapy.

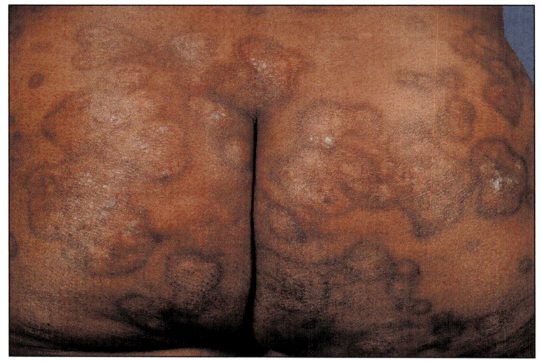

Annular psoriatic plaques of buttocks

In most sufferers, the tendency to get psoriasis is genetic. Inherited psoriasis usually starts in older childhood or as a young adult. Sometimes, especially in children, a virus or strep throat triggers brief attacks of tiny spots of psoriasis. In middle-aged adults, a non-hereditary type of psoriasis can develop. This changes more rapidly than the inherited form, varying in how much skin is involved more unpredictably. Most types of psoriasis show some tendency to come and go, with variable intensity over time. Medications may trigger a flare up weeks to months after starting them. These include non-steroidal anti-inflammatory drugs, beta-blockers, oral steroids, and antidepressants.

In cases of inverse psoriasis in sensitive skin areas, (i.e. eyelids, ears, flexures, sub-mammary, axillae, anal fold) and in children, low potency topical corticosteroids may be appropriate. High potency corticosteroids may be used for treatment resistant plaques, particularly those on the hands or soles. Except for very rare circumstances, systemic steroids are contraindicated for the treatment of psoriasis. There is a possibility of a rebound effect. The rebound effect can also occur with topical corticosteroid use of high and super high groups with occlusion.

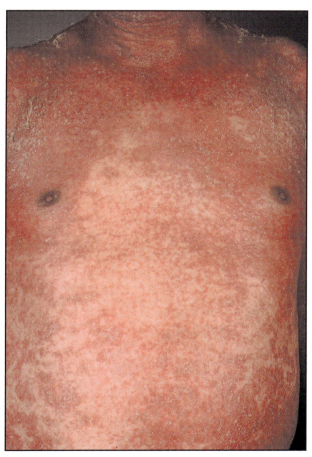

Psoriasis with resulting exfoliative erythroderma

Potential Side Effects

Like all kinds of medications, topical corticosteroids have the potential to pro-
duce adverse reactions. Side effects of topical corticosteroids limit their long-term
use. Prolonged use of topical corticosteroids and systemic immunosuppressant
drugs (i.e., corticosteroids, cyclosporine, azathioprine) can result in severe cutaneous and systemic effects. Some corticosteroids, especially the nonfluorinated
double-ester type such as prednicarbate, seem to affect fibroblast growth in vitro
as well as skin thickness in vivo less than equipotent conventional corticosteroids.

Cutaneous atrophy is the most common adverse effect. Atrophy can occur when
potent steroids are applied for several months, especially to areas where skin is

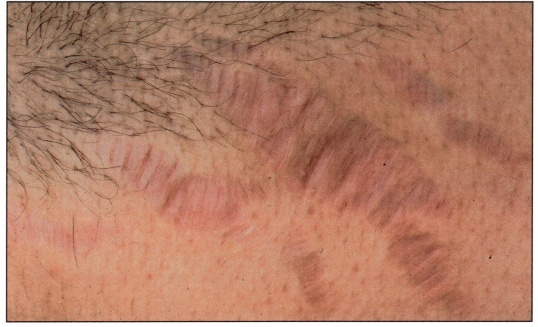

Striae secondary to high potency steroid use

thin already such as the flexures. Tachyphylaxis is a common effect of chronic topical corticosteroid application. Rebound dermatitis occurs in some skin disorders when the topical steroid is discontinued. It is especially common in psoriasis and rosacea patients. Perioral dermatitis is frequently seen on the face after the

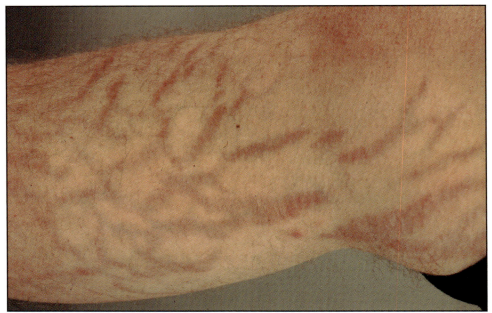

Striae secondary to high potency steroid use

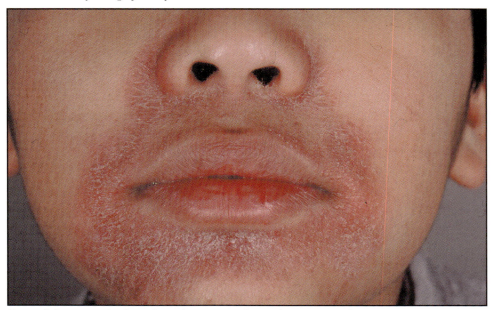

Perioral dermatitis resulting from chronic use of topical corticosteroids

use of topical corticosteroids, although it may also result from the use of moisturizers. Erythematous, irritable, and slightly scaly papules arise periorally, sparing the vermilion border. A similar rash may be seen around the nostrils and eyelids. The use of potent topical corticosteroids applied to the breasts, abdomen, thighs,

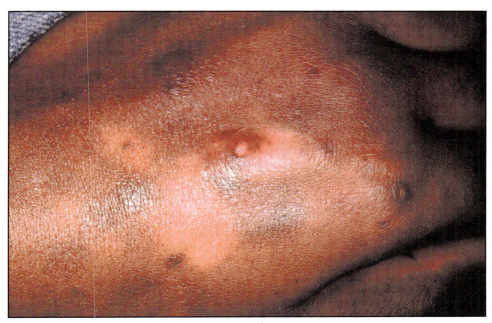

Hypopigmentation resulting from long term steroid use

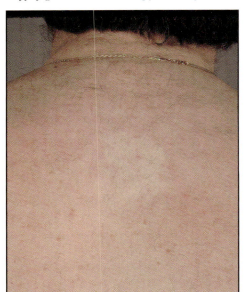

Localized hyopoigmented patch resulting from long term steroid use

legs and upper arms for more than several weeks may cause striae due to the breakdown of collagen fibers. Pruritis may result from long-term application of topical steroids, and is especially common if the skin is itchy.

The level of contact sensitivity can be mild or severe, and some patients who demonstrate mild patch test responses may be allergic to multiple corticosteroids. Susceptibility to systemic effects of topical steroid use varies. Local adverse effects occur more frequently than systemic effects. The risk with unoccluded topical corticosteroids is significantly lowered.

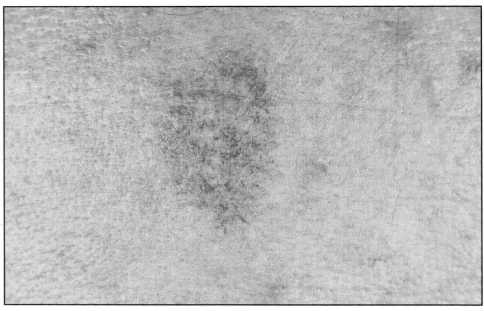

Telangiectasias resulting from long term steroid use

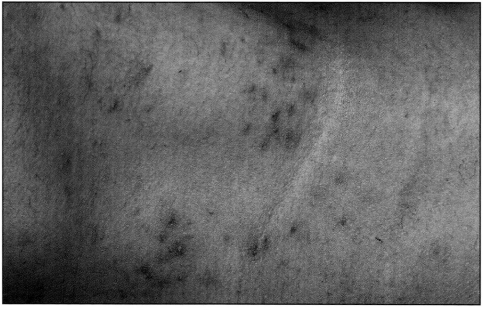

Occlusion folliculitis secondary to use of topical corticosteroids

TABLE VII

COMMON SIDE EFFECTS OF TOPICAL CORTICOSTEROID USE

CUTANEOUS SKIN CHANGES

Skin blanching from acute vasoconstriction

Hypopigmentation

Rebound worsening of the pre-existing skin condition

Miliaria

Rosacea, perioral dermatitis, acne

Skin atrophy with telangiectasia, stellate pseudoscars, purpura, striae

Delayed wound healing

Hypertrichosis of face

Allergic contact dermatitis

CUTANEOUS INFECTION AND INFESTATION

Folliculitis

Tinea incognito

Impetigo incognito

Scabies incognito

EYES

Glaucoma

Cataracts

Systemic

Adrenal suppression

Osteoporosis

Stunted growth in children

Cushingoid appearance

7

Summary & Conclusion

Topical corticosteroids have long represented the only highly effective weapon Dermatologists had in the treatment of dermatoses. Antibiotics, antihistamines, and other therapies have also been used, but their benefits are modest and often limited in duration. However, recent clinical trial data have demonstrated the efficacy, convenience, and long-term safety of new topical non-steroid drugs that have very encouraging side effect profiles. In the near future, the spectrum of treatment options available will be able to meet both patients' and physicians' needs for long-term management of symptoms and will be effective in treating treat all affected sites, including the face.

Resources

www.eczema-assn.org

www.eczema.org

www.psoriasis.org

www.aad.org

www.psoriasis.org/npf.shtml

www.eczemainformant.com

www.pedeczema.com

www.allallergy.net

References

Staughton, Richard. Psychologic approach to atopic skin disease. Journal of the American Academy of Dermatology 2001; 45:S53.

Brehler R, Hildebrand A, Luger T. Recent developments in the treatment of atopic eczema. Journal of the American Academy of Dermatology 1997; 36:983-994.

Clark AR, Jorizzo JL, Fleischer AB. Papular dermatitis (subacute prurigo, "itchy red bump" disease): pilot study of phototherapy. J Am Acad Dermatol 1998; 38:929-33.

Taieb, Alain. The natural history of atopic dermatitis. Journal of the American Academy of Dermatology 2001; 45:S4-S6.

Tofte Susan J, Hanifin Jon M. Current management and therapy of atopic dermatitis. Journal of the American Academy of Dermatology 2001; 44:S13-S16.

Krueger G, et al. Two considerations for patients with psoriasis and their clinicians: What defines mild, moderate, and severe psoriasis? What constitutes a clinically significant improvement when treating psoriasis? Journal of the American Academy of Dermatology 2000; 43:281-285.

Koo John, Lebwohl Mark. Duration of remission of psoriasis therapies. Journal of the American Academy of Dermatology 1999; 41:51-59.

Hiratsuka Sachie, Yoshida Akira, Ishioka Chihiro, Kimata Hajime. Enhancement of in vitro spontaneous IgE production by topical steroids in patients with atopic dermatitis. Journal of Allergy and Clinical Immunology 1996; 1:107-113.

Nghiem Paul, Pearson Greg, and Langley Richard. Tacrolimus and pimecrolimus: From clever prokaryotes to inhibiting calcineurin and treating atopic dermatitis. Journal of the American Academy of Dermatology 2002; 46:228-241.

Grundmann-Kollmann Marcella, Behrens Stefanie, Podda Maurizio, Peter Ralf Uwe, Kaufmann Roland, Kerscher Martina. Phototherapy for atopic eczema with narrow-band UVB. Journal of the American Academy of Dermatology 1999; 40:995-997.

Shin Helen T, Chang Mary Wu. Drug eruptions in children. Current Problems in Pediatric and Adolescent Health Care 2001; 31:207-215.

Boguniewicz M, Fielder VC, Raimer S, Lawrence ID, Leung DYM, Hanifin JM. A randomized vehicle controlled trial of tacrolimus ointment for treatment of atopic dermatitis in children. J Allergy Clin Immunol 1998; 102:635-42.

Hanifin Jon, Chan Sai. Biochemical and immunologic mechanisms in atopic dermatitis: New targets for emerging therapies. Journal of the American Academy of Dermatology 1999; 41.

Barnes Peter J. New directions in allergic diseases: Mechanism-based anti-inflammatory therapies. Journal of Allergy and Clinical Immunology 2000; 106.

Boschert Sherry. Strategies for Tough-to-Treat Pediatric Eczema. Skin and Allergy News 2001; 32(12):34.

Gilbertson Erik O, Spellman Mary C, Piacquadio Daniel J, Mulford Mim I. Super potent topical corticosteroid use associated with adrenal suppression: Clinical considerations. Journal of the American Academy of Dermatology, February 1998 (2); 38(2):318-21.

Al-Suwaidan Sami N, Feldman Steven R. Clearance is not a realistic expectation of psoriasis treatment. Journal of the American Academy of Dermatology, May 2000 (1): 42(5):796-802.

Stern RS. The pattern of topical corticosteroid prescribing in the United States, 1989-1991. Journal of the American Academy of Dermatology, August 1996; 35(2):183-6.

Leung DYM. Atopic dermatitis and the immune system: The role of superantigens and bacteria. Journal of the American Academy of Dermatology, July 2001(2); Volume 45(1):S13-S16.

Guin JD. Contact sensitivity to topical corticosteroids. Journal of the American Academy of Dermatology, May 1984; 10(5):773-82.

Finlay Andrew Y. Quality of life in atopic dermatitis. Journal of the American Academy of Dermatology, July 2001(2); 45(1):S64-S66.

Tofte Susan J, Hanifin Jon M. Current management and therapy of atopic dermatitis. Journal of the American Academy of Dermatology, January 2001(2); 44(1):S13-S16.

Lebwohl Mark, Ali Suad. Treatment of Psoriasis, Part 1: Topical therapy and phototherapy. Journal of the American Academy of Dermatology, October 2001; 45(4):487-498.

Thompson, EB. The Structure of the Human Glucocorticoid Receptor and its Gene. Journal of Steroid Biochemistry 1987; 27:105-108.

Hanifin JM, Tofte SJ. Update on therapy of atopic dermatitis. J Allergy Clin Immunol 1999; 104:123-125.

Ruzicka T, Ring J, Przybilla B. Handbook of Atopic Eczema. Berlin, Springer Verlag, 1992.

Rajka G. Essential Aspects of Atopic Eczema. Berlin, Springer Verlag, 1990.

Ring J. Atopic diseases and mediators. Int Arch Allergy Immunol 1993; 101:305-307.

Leung DYM. Immune mechanisms in atopic dermatitis and relevance to treatment. Allergy Proc 1991; 12:339-346.

Kapp A. Atopic dermatitis-the skin manifestation of atopy. Clin Exp Allergy 1995; 25:210-219.

White A, Horne DJL, Varigos GA. Psychological profile of the atopic eczema patient. J Dermatol 1990; 31:13-16.

Koblenzer C. Stress and the skin: Psychosomatic concepts in dermatology. Arch Dermatol 1983; 119:501-512.

King RM, Wilson GV. Use of a diary technique to investigate psychosomatic relations in atopic dermatitis. J Psychosom Res 1991; 35:697-706.

Mommaas AM. Influence of glucocorticoids on the epidermal Langerhans cell. Curr Probl Dermatol 1993; 21:67-72.

Boguniewicz M, Leung DYM. Management of atopic dermatitis. In: Leung DYM, ed. Atopic dermatitis: from pathogenesis to treatment. Austin, Tex.: RG Landes, 1996:185-220.

Notes

Notes

Notes

Notes